Cyber Mission Thread Analysis

A Prototype Framework for Assessing Impact to Missions from Cyber Attacks to Weapon Systems

DON SNYDER, ELIZABETH BODINE-BARON, DAHLIA ANNE GOLDFELD,
BERNARD FOX, MYRON HURA, MAHYAR A. AMOUZEGAR,
LAUREN KENDRICK

Prepared for the Department of the Air Force
Approved for public release; distribution unlimited

PROJECT AIR FORCE

For more information on this publication, visit **www.rand.org/t/RR3188z1**.

About RAND

The RAND Corporation is a research organization that develops solutions to public policy challenges to help make communities throughout the world safer and more secure, healthier and more prosperous. RAND is nonprofit, nonpartisan, and committed to the public interest. To learn more about RAND, visit www.rand.org.

Research Integrity

Our mission to help improve policy and decisionmaking through research and analysis is enabled through our core values of quality and objectivity and our unwavering commitment to the highest level of integrity and ethical behavior. To help ensure our research and analysis are rigorous, objective, and nonpartisan, we subject our research publications to a robust and exacting quality-assurance process; avoid both the appearance and reality of financial and other conflicts of interest through staff training, project screening, and a policy of mandatory disclosure; and pursue transparency in our research engagements through our commitment to the open publication of our research findings and recommendations, disclosure of the source of funding of published research, and policies to ensure intellectual independence. For more information, visit www.rand.org/about/principles.

RAND's publications do not necessarily reflect the opinions of its research clients and sponsors.

Preface

This report presents the results of a project commissioned by the Commander of the Air Force Life Cycle Management Center and the Deputy Assistant Secretary of the Air Force for Science, Technology, and Engineering, Office of the Assistant Secretary of the Air Force for Acquisition and Logistics, to develop an improved process for assessing the cybersecurity risk of weapon systems that incorporates mission impact. The process is designed to be useful within the acquisition community and specifically to address the impact of cyber attacks to the missions these systems support. The ultimate goal is to improve decisionmaking when setting cybersecurity risk mitigation priorities throughout the life cycle of weapon systems. The work contributed to Line of Action #1 of the U.S. Air Force Cyber Campaign Plan. It has been modified based on insights from some pilot applications and we expect that it will further evolve as additional lessons are learned. It might differ in detail from the methodology adopted by the U.S. Air Force.

This research builds on previous work for the life-cycle management community on cybersecurity in life-cycle management.[1] It was initiated as part of a fiscal year 2015 project, "Enhancing Cyber Mission Assurance Through Improved Measurement of Effectiveness," and revised and completed as part of a fiscal year 2017 project, "Refinement of Cybersecurity Risk Assessment Framework and Cyber Mission Thread Analysis." The research reported here was conducted within the Resource Management Program of RAND Project AIR FORCE. It should be of interest to the cybersecurity, acquisition, test, and operational communities.

RAND Project AIR FORCE

RAND Project AIR FORCE (PAF), a division of the RAND Corporation, is the Department of the Air Force's (DAF's) federally funded research and development center for studies and analyses, supporting both the United States Air Force and the United States Space Force. PAF provides the DAF with independent analyses of policy alternatives affecting the development, employment, combat readiness, and support of current and future air, space, and cyber forces. Research is conducted in four programs: Strategy and Doctrine; Force Modernization and Employment; Resource Management; and Workforce, Development, and Health. The research reported here was completed under contract FA7014-16-D-1000.

Additional information about PAF is available on our website:
www.rand.org/paf

---

[1] Don Snyder, James D. Powers, Elizabeth Bodine-Baron, Bernard Fox, Lauren Kendrick, and Michael H. Powell, *Improving the Cybersecurity of U.S. Air Force Military Systems Throughout Their Life Cycles*, Santa Monica, Calif.: RAND Corporation, RR-1007-AF, 2015.

This report documents work originally shared with the U.S. Air Force on May 30, 2018. The draft report, issued on October 13, 2017, was reviewed by formal peer reviewers and U.S. Air Force subject-matter experts.

Contents

Preface iii

Figures vi

Tables vii

Summary viii

Acknowledgments x

Abbreviations xi

1. Some Considerations for Assessing Cybersecurity Risk of Weapon Systems 1

　　Introduction 1

　　Why Another Methodology? 3

　　Is the Risk from Cyber Attacks Different from Other Risks to Missions? 5

　　The Practice of Cyber Mission Thread Analysis 6

2. A Prototype Framework for Assessing Impact to Mission 7

　　Overview 7

　　Defining a Mission 9

　　Triage 14

　　Summary 31

3. Discussion of the Framework 33

　　How This Method Addresses the Key Challenges 33

　　Metrics for Cybersecurity 34

　　Discussion and Conclusions 35

References 38

Figures

Figure 2.1. Overview of the Framework 8

Figure 2.2. Modularity of Cyber Mission Threads...13

Figure 2.3. Sample Functional Flow Block Diagram..17

Figure 2.4. Linear Mission Example 19

Figure 2.5. Linear Mission with Alternative Paths Example 19

Figure 2.6. Notional Mission Element Graph 21

Figure 2.7. Boolean Logic Transformation from Functional Flow Diagram to Graph.................22

Figure 2.8. Conditional Logic Transformation from Functional Flow Diagram to Graph 23

Figure 2.9. Alternative Conditional Logic Transformation from Functional Flow Diagram
to Graph...23

Figure 2.10. Depiction of Cyber Separability of Systems Supporting a Critical Function...........28

Figure 2.11. Generalization of the Failure and Recovery of a Capability After an Attack...........31

Tables

Table 2.1. Possible U.S. Air Force Missions for Cyber Mission Thread Analysis 12

Table 2.2. Cut Sets and Criticality Tiers for a Notional Mission 21

Summary

There is an increasing need to understand the risks to missions from cyber attacks against weapon systems. RAND Corporation was asked to develop a framework for mission impacts to guide decisions of where to mitigate and where to accept risk by authorizing officials[1] and program offices. Previous frameworks struggled with four dimensions of the problem:

- The need to be executable at the scale of all Air Force missions
- The need to be updated throughout the life cycle of a weapon system as concepts of operations change, as new systems are introduced and legacy systems modified, and as the threat evolves
- The complexity of assessing criticality to missions if criticality depends on the scenario
- The need to be transparent so that decisionmakers understand enough about how the analysis works and its limitations to trust it to guide decisions.[2]

A further goal is to address two peculiarities of cybersecurity assessments that differ from other mission impact assessments:

- In the cyber realm, redundancy does not provide robustness.
- Loss of command and control can injure a mission without any system or component failure.

The prototype framework presented in this report views risk as a combination of vulnerabilities to systems, threats exploiting those vulnerabilities, and the eventual impact to the mission(s) if the threat is realized. The strong emphasis is on making the mission impact assessment tractable in the face of the four challenges and two peculiarities of cyberspace. Detailed vulnerability and threat assessments can be done later using this methodology to guide where to focus analytical effort based on where the mission is most affected.

In analyzing mission impact, the methodology follows a top-down approach, from the overall mission through mission elements to systems. Care needs to be taken to define missions that are not too broad to assemble an appropriate working group for assessment and not too narrow to cause duplication of effort and gaps in analysis. After a mission is selected and defined, the next step is to decompose the mission into mission elements, then decompose the systems that support

---

[1] An *authorizing official* is an official empowered to make risk-based decisions regarding the acceptance of cybersecurity risk of operating systems that the Department of Defense (DoD) uses to process, store, or transmit information. In this capacity, authorizing officials report to the chief information officer (CIO) and assist in carrying out the CIO's statutory responsibilities. Authorizing officials replace the former designated approving or accrediting authorities. See Department of Defense, *Cybersecurity*, Instruction 8500.01, March 14, 2014.

[2] We use the term *criticality* to mean the quality of being a key system or mission element whose failure would place a mission at risk.

those missions. The functional flow among the mission elements is also defined in this step. The methods employed can be drawn from standard methods used in systems engineering. The idea is to fully describe the mission elements, the supporting systems, and their functional flow.

For assessing the relative mission impact of cyber attacks, it is useful to rank the relative criticality of the mission elements, or at least group them into sets of relative criticality. We propose a topological method for the first ranking and triage. The method described is based on cut set analysis from graph theory.[3] Following this first triage, we propose four additional triaging steps based on system attributes:

1. The degree of system dependencies. Dependencies of systems on one another provide a measure of fragility of the architecture.
2. A concept we call *cyber distance*, which is the minimum number of hops of different protocols between the system of interest and a nonsecure network. Cyber distance provides a proxy for how accessible a system is in cyberspace and serves as an approximation of vulnerability.
3. A new concept we call *cyber separability*. Two systems are cyber separable from one another if no single cyber attack vector could simultaneously degrade the functionality of both. Cyber separability is a proxy metric for robustness to a cyber attack.
4. The relative (statistical) timing of an attack versus the (statistical) timing of a recovery, which provides a measure of whether the adversary or the defender has the relative advantage. Although not always easy to assess, when it is possible, the relative timing provides a powerful insight into the mission impact.

These criticality criteria also form the foundation for defining metrics of cybersecurity at the system (or program) level.

After any system-specific pruning, the stage is reached for detailed assessment of vulnerabilities and threats, possible mitigations, and costs of implementing those implementations. The program manager and authorizing official would use these insights to guide which risks might be accepted and which should be explored for mitigations. Those mitigations might be found in a number of areas: by changing the functional flow logic at the mission level, by changing the concepts of operations at a more detailed level, or by making a change to a system or system interfaces.

There is no transcendentally correct method for assessing mission impact or cybersecurity risk. The method we present in this report is one approach developed to satisfy the constraints and desirable characteristics described above. To establish that we have adequately met those goals, and to fully verify and validate the framework, the framework will need to be applied to several different missions. During that process, it is inevitable that insights into the framework and methodologies will be gained and they will need to be refined.

---

[3] Graph theory is a branch of mathematics on the study of networks. Cut set analysis is one technique within graph theory for determining criticality of nodes or links within a network.

Acknowledgments

We thank Lt Gen John F. Thompson for initially sponsoring this work and for championing it throughout. In fiscal year 2017, Lt Gen Robert D. McMurry, Jr., and Jeffrey Stanley continued support of the project, for which we are grateful. We are very appreciative of all the work that Dennis Miller, Mitchell Miller, and Danny Holtzman have done to assist and promote the execution of this work. It could not have been done otherwise. We thank Col William Young for many helpful discussions on mission-level analysis. Many others in the Air Force, at MITRE, and at Booz Allen Hamilton also helped and supported us. While they are too numerous to mention by name, their ideas and critiques were indispensable.

At RAND, we thank James Chow, John G. Drew, Lionel Galway, Lara Schmidt, and Guy Weichenberg for helpful discussions.

Formal reviews by Lionel Galway, Andrew Lauland, and Isaac Porche improved the document.

That we received help and insights from those acknowledged here should not be taken to imply that they concur with the views expressed in this report. We alone are responsible for the content, including any errors or oversights.

Abbreviations

CIO	chief information officer
CRRA	Capabilities Review and Risk Assessment
CVSS	Common Vulnerability Scoring System
DoD	Department of Defense
IT	information technology
NIPRNet	Nonsecure Internet Protocol Router Network
OODA	observe, orient, decide, act
PAF	Project AIR FORCE
STPA	System Theoretic Process Analysis
STPA-Sec	System Theoretic Process Analysis for Security

1. Some Considerations for Assessing Cybersecurity Risk of Weapon Systems

Introduction

The most important consideration for whether to mitigate or accept a risk from a cyber attack is how it affects operational missions. Yet weaving the thread from a potential cyber attack to what impacts it might have on missions has proved to be a difficult task. This report presents a prototype framework for assessing the mission impact component of cybersecurity risk specifically aimed to be executable at the scale of a comprehensive assessment of all missions across the entire Air Force. The framework is motivated by assessing mission impact of cyber attacks against weapon systems, but these concepts are generally extensible to information technology (IT) systems.

Exploitation in the form of intelligence collection through cyberspace, such as the theft of design drawings and other sensitive program information, either from the government or its industrial base, is outside the scope of this effort. The loss of sensitive program information can, of course, have a mission impact. The adversary can use the information to develop more effective tactics against U.S. systems and use the technology to develop better systems. However, espionage and direct cyber attack are quite different adversary activities and require separate frameworks for analysis of mission impacts.

Our goal is to provide a decisionmaking aid for authorizing officials[1] and system program offices that helps them determine when to further investigate vulnerabilities and threats and informs them where to accept and where to mitigate risk from cyber attack—a method simple enough to perform in no more than a few months, but informative enough for decision support.

We define *cybersecurity risk* as the possibility of operational harm caused by an adversary's action through cyberspace. Our premise is that the possibility of harm can be specified as a combination of vulnerabilities of systems, threats to those systems, and the impacts to the operational missions those systems support if the threats are realized. We define these elements more precisely here as the terms are not always used the same way within the Department of Defense (DoD).[2]

---

[1] An *authorizing official* is an official empowered to make risk-based decisions regarding the acceptance of cybersecurity risk of DoD operating systems that process, store, or transmit information. In this capacity, authorizing officials report to the chief information officer (CIO) and assist in carrying out the CIO's statutory responsibilities. Authorizing officials replace the former designated approving or accrediting authorities. See Department of Defense, March 14, 2014.

[2] See Yacov Y. Haimes, "On the Complex Definition of Risk: A Systems-Based Approach," *Risk Analysis*, Vol. 29, No. 12, 2009, pp. 1647–1654, and Terje Aven, "The Risk Concept—Historical and Recent Development Trends," *Reliability Engineering and System Safety*, Vol. 99, 2012, pp. 33–44, for a recent review.

Vulnerabilities are weaknesses in the design or configuration of a system, or how that system is used, that enable an attacker.[3] A simple example is software that allows a buffer overflow. If an adversary were able to gain access to that software because it is connected to the internet (or by other means), the adversary might be able to exploit the buffer overflow to execute malware. The combination of the undesirable configuration (buffer overflow) and relatively easy access (connection to the internet) would be a significant vulnerability. On the other hand, software that cannot easily be placed in an undesirable configuration and that resides on an air-gapped system would be assessed as having lower vulnerability. Clearly, the more complex a system, the less confident the owner can be in understanding all possible configurations and hence of all possible vulnerabilities. Most modern electronic systems are complex and, because of the need for information flow in modern systems, no system is fully isolated. How individuals interact with a system is also difficult to fully control. Any modern system of high complexity that processes and transfers information most likely has at least moderate vulnerability to some form of cyber attack.

Threats are the capabilities that adversaries possess (or are assessed to possess in the future) and their intent to use those capabilities against the United States. The cyber threat depends on the adversary (capabilities and intent) and the scenario (which adversaries and their intentions). Numerous adversaries pose threats to both systems and the missions that they support. Given the wide range of potential adversaries and missions, and the vast number of systems supporting those missions, comprehensive threat assessments are impractical. Beyond the large scope, threat assessments are difficult in the cyber domain because of the ever-changing cyber environment. In addition to the general challenges of threat assessment, cyber technologies rapidly evolve. It is not clear that potential adversaries have shown their full capabilities, and even they might not know what their future intentions might be to use their capabilities. Assessed threats prove challenging to express in terms that can be applied easily to a specific assessed vulnerability.

Mission impact is how much the performance of an operational mission is diminished by an exploited vulnerability in a supporting system. A mission is carried out by (generally) a large number of materiel and non-materiel elements that work in concert. Any given system is just one enabling part of a given mission, and most systems enable more than one mission. Furthermore, in order to be robust, most missions can be carried out by more than one specific combination of materiel and non-materiel elements. Exactly what combination of elements might be put into service for a mission can also evolve over time as existing systems are modified, new systems are introduced, and concepts of operations change. Different operating locations might require

---

[3] There is no single definition of *vulnerability* in the DoD. This definition is consistent with those adopted by Committee on National Security Systems, *Committee on National Security Systems (CNSS) Glossary*, CNSSI No. 4009, April 6, 2015, and the Joint Chiefs of Staff, *DOD Dictionary of Military and Associated Terms*, February 2019. Some use the term *susceptibility* to mean more or less the same thing (e.g., Mark M. Stephenson, *Avionics Cyber Vulnerability Assessment and Mitigation Manual*, Air Force Research Laboratory, March 2014. Not available to the general public). For a discussion of definitions of vulnerability, see also Yacov Y. Haimes, "On the Definition of Vulnerabilities in Measuring Risks to Infrastructures," *Risk Analysis*, Vol. 26, No. 2, 2006, pp. 293–296.

different resources to perform a particular mission. The connections among systems and missions are, therefore, complex and constantly evolving.

For the purposes of risk assessment, vulnerabilities are attributes of systems and the procedures in which they are used. Threats are an attribute of the environment. And mission impact is the effect on operations.

To decide where to accept risk and where to invest resources to reduce risk, decisionmakers need three inputs: (1) an overall assessment of risk, being a combination of vulnerabilities, threats, and mission impact; (2) potential mitigations for those risks; and (3) the costs for implementing those mitigations. This report focuses on the risk part of this decision support sequence. But given the sheer number of systems to be analyzed throughout the U.S. Air Force, it is futile to attempt to assess all of them for vulnerabilities and threats. We take it as a premise that some risk must be accepted, at least in the short term, and that risk should be accepted in areas of low to moderate mission impact.

It is plausible, as we argue in this report, to assess which systems have the greatest overall impact to a mission if they were to fail because of a cyber attack. This mission impact assessment can be used as a triage to identify a selected number of mission elements and systems for more detailed analysis of vulnerabilities and threats. Doing a mission impact analysis as the first step in a risk assessment thereby reduces the workload to a manageable number of vulnerabilities and threats to assess.

Because the mission impact assessment is meant to triage which systems and mission elements to examine in more detail for vulnerabilities and threats (and, ultimately, for mitigations and costs), the omission of a critical system or mission element from the final list of critical items is of more concern than the inclusion of a system or mission element that should not be on the list.[4] Some false positives are okay. When a potentially false positive appears, further examination should identify it as a potential false positive without much resource expenditure. The reason why such a system or mission element ended up on the list might itself be informative. In one recent case in a pilot application of this methodology, a system was identified as critical that was rapidly judged to be unworthy of further analysis of vulnerability or threat. The system appeared on the list because several processes routinely used that system for convenience rather than the system of record to perform some critical mission elements. The initial identification of that system as critical (when it was not) revealed lapses in how processes are actually done relative to how they are supposed to be done.

Why Another Methodology?

Although the risk analysis literature overflows with methodologies for assessing risk and the impacts to missions, most of them are not designed for assessing impacts to missions at the scale

---

[4] We use the related term *criticality* to mean the quality of being a key system or mission element whose failure would place a mission at risk.

of the breadth of U.S. Air Force missions. The main objective of this report is to incorporate mission impact into risk assessments for cybersecurity in a way that can be executed at scale across the U.S. Air Force. Previous attempts to do so have proven problematic because of several challenges; this new proposed methodology aims to at least partially address those challenges.

The first is one of scale. Even a narrowly defined mission requires a vast number of systems. These are not unique; there are often multiple systems that can perform a part of a mission (e.g., KC-10 or KC-135 for tactical refueling), and each of these can perform parts of multiple missions (e.g., the aerial refuelers can also transport cargo and evacuate medical patients). Each system can be quite complex, leading to an enormous number of potential vulnerabilities to examine. And each of these vulnerabilities can have a complicated threat environment that depends on the scenario and the adversary, including projections of these capabilities and intentions well into the future when considering a new program whose life span might be measured by decades. That is just considering a single, narrowly defined mission. Multiply the task across every mission in the U.S. Air Force and the number of elements to assess explodes. A need exists for managing this scale so that analyses can be executed with reasonable resource expenditures.

The second challenge is one of assessment life span. How a mission is performed changes as new systems are introduced, legacy systems are modified, and concepts of operations, and tactics, techniques, and procedures evolve. Changes to the systems induce changes to the system's vulnerabilities, and concurrently, the threat evolves. Consequently, a need exists for reexamining the risk assessment as the missions, vulnerabilities, and threats change. Nevertheless, some core insights should emerge from the assessment that are not so ephemeral because acquisition decisions with long-term ramifications need to be made.

The third challenge is that what is deemed critical to a mission depends on the scenario in which the mission is to be performed. Which missions are most important and how much degradation of a mission is possible and yet still achieve the commander's intent are situationally dependent. There are a lot of scenarios that most missions can support, ranging over location and intensity (e.g., counterterrorism and arresting the incursions of a near-peer adversary). Any method that focuses on just one of these scenarios does not serve the decisionmaker well.

The fourth challenge is one of transparency and confidence. The more opaque a methodology is, the more it is perceived as a "black box" and therefore less trustworthy. Decisionmakers will often revert to intuition and professional judgment when they do not understand the basic assumptions and workings of analysis. This reaction presents a motivation to be transparent, use existing and proven techniques when possible, and be explicit about assumptions and uncertainties.

An excellent case study where most of these problems surfaced was the Capabilities Review and Risk Assessment (CRRA), which was the last time the U.S. Air Force attempted to

4

systematically assess mission impact across the service.[5] The CRRA was an initiative that started in 2002 and lasted until about 2011. Its goal was to prioritize budgetary decisions based on mission impact. In the first years, it was performed annually and later every two years. Each time, the methodology was somewhat adjusted, but throughout, it essentially decomposed missions into comprehensive, mutually exclusive elements until a level of detail was reached to link mission impacts to funding for program elements.

CRRA participants struggled with the vastness of the challenge and often stopped the analysis at a level of detail too coarse to address meaningful issues. The higher level of detail sought, the greater the need for specific subject-matter expertise, and such expertise was often not available. Although each CRRA took one to two years to perform, the CRRA did not by design encounter significant issues with shifting missions, vulnerabilities, and threats over this period. The reason was that a particular date in the future was used for the analysis baseline, and how missions would be performed and what systems would be used was defined against that date. However, this choice led to uncertainties of what concepts of operations, systems, and threats might exist at that future date, and all participants did not choose consistently. For assessing mission criticality, each CRRA generally used as a baseline a single scenario drawn from defense planning scenarios. Some programs were then considered a low priority, whereas they might have been assessed as a high priority had another scenario been selected. Which scenario was used as a baseline also sometimes varied by work group, as there is an incentive to avoid a scenario to assess a mission in which that mission plays a trivial role.[6] Finally, the methodologies employed were executed differently by different working groups and not clearly described to senior leaders, who then lacked confidence in the outcomes.[7]

Is the Risk from Cyber Attacks Different from Other Risks to Missions?

For the most part, an attack through cyberspace or a routine failure of a component caused by a cyber bug is no different from any other failure mode. What matters is the effect on the system and mission, not how the effect happened. So the question arises: From the perspective of mission impact, is there anything different about cyber risk assessment that is not captured in a generic risk assessment? We argue that there are two aspects of the cyber problem that present qualitatively different dimensions that must be addressed.

First, unlike virtually all other failure modes, redundancy does not contribute to robustness against cyber attack. Redundant components, by which we mean components that are essentially

---

[5] See Don Snyder, Patrick Mills, Adam C. Resnick, and Brent D. Fulton, *Assessing Capabilities and Risks in Air Force Programming: Framework, Metrics, and Methods*, Santa Monica, Calif.: RAND Corporation, MG-815-AF, 2009, for a brief description and assessment of the CRRA.

[6] For example, any single scenario strains to serve as a good baseline for both nuclear and the full range of nonnuclear missions.

[7] These observations are based on the senior author's experience in working with the CRRA during its tenure.

identical and designed as backups, share a common vulnerability to cyber attack and hence do not provide robustness. If an attacker can install malware to erase one hard drive, it can probably also erase a backup hard drive. We develop this concept fully in Chapter 2.

Second, cyber effects can happen that are not component failures. If an adversary takes command and control of a system, all of the components of that system might act within the bounds of acceptable performance but at a time or place of the adversary's choosing that is injurious to the mission. This kind of behavior more resembles a disgruntled insider threat and is not normally captured in techniques used in systems engineering for safety.

Therefore, some adjustments to traditional methods are required to include the full range of cyber effects. Nevertheless, given that cyber threats share much in common with other threats, it would be an imprudent use of resources to develop a separate methodology for assessing cybersecurity risks and risks caused by other factors. What we propose in this report is a combination of elements selected from accepted methods that combine to assess mission impact.

The Practice of Cyber Mission Thread Analysis

The prototype framework in this report was developed to address the problem of assessing mission impact from cyber attack *comprehensively across the U.S. Air Force*. Detailed assessments of every system and mission in the Air Force are not practical. For that reason, we will offer a method for triaging the analysis and approximating attributes with proxies. It is formulated to be done at scale, to be amenable to being updated as needed, to be applicable across scenarios, and to be clear in how it works. It also specifically addresses the peculiar characteristics of cyberspace: the ineffectiveness of redundancy and the vulnerability to adversary manipulation of command and control. In short, this framework is meant to be comprehensive enough that it can be executed at the scale of each of the missions in the U.S. Air Force yet simple enough that it can be executed and updated as needed.

A previous draft of this methodology was used in a cyber mission thread analysis of the aerial refueling mission as part of Line of Action #1 of the U.S. Air Force Cyber Campaign Plan. Based on that experience, and to a lesser extent the experience of a second cyber mission thread analysis of the U.S. Air Force nuclear command, control, and communications mission, we have revised some of the details of the methodology. We expect that it will be further refined as additional experience is gained. Given this evolution, the methodology presented here should not be viewed as the official U.S. Air Force methodology.

In a companion document, we propose detailed steps for implementing the methodology presented in this report.[8]

---

[8] Lauren A. Mayer, Don Snyder, Guy Weichenberg, Danielle C. Tarraf, Jonathan W. Welburn, Suzanne Genc, Myron Hura, and Bernard Fox, *Cyber Mission Thread Analysis: An Implementation Guide for Process Planning and Execution*, Santa Monica, Calif.: RAND Corporation, RR-3188/2-AF, 2022.

2. A Prototype Framework for Assessing Impact to Mission

In this chapter, we describe the prototype framework for cyber mission thread analysis and criticality identification. For the first few times this methodology is executed, we expect that further refinements will be needed. Although developed with cybersecurity in mind, the framework is meant to be general enough to use for examining mission impact during broader risk assessments so as to not require two separate processes. We begin with a brief overview here to orient the reader for the later sections, which will describe the framework in more detail. Because the emphasis is on mission impact, the approach starts with missions not systems, vulnerabilities, or threats.

Overview

The methodology followed throughout is one that works from the top-down, from the overall mission through mission elements to systems.[1] In contrast, a bottom-up approach starts at a high level of detail with system components and works upward to the overall U.S. Air Force missions. A bottom-up approach requires exhaustively identifying and analyzing every component of every system and every interaction among those systems, then combining them into a system-of-systems that performs one or more missions. While in theory this would allow us to trace how failure or compromise of one system proliferates into mission impacts, the combinatorial explosion of effort required prohibits a bottom-up approach for assessing the entire portfolio of missions.

The first step is to select and define a mission, followed by decomposing that mission into mission elements. At this stage, the emphasis is on a mission and is as much as possible agnostic of specific systems. The functional flow among the mission elements will also be defined by the user in this step. The decomposition methods can be drawn from standard methods used in systems engineering but applied in this case to mission elements rather than to system components. The idea is to express what activities need to be done in the mission and not how those activities are done or what systems might be employed to do them. This step is depicted by the top box (functional-level analysis) in Figure 2.1.

However, initial experience has indicated that mission elements and systems cannot be cleanly separated. Even at the highest level, some mission elements are intertwined with systems in a way that makes them difficult to disentangle. One example is the activity of a platform communicating with a weapon that it carries. We can abstract this to a mission element of passing certain critical data between the platform and the weapon. But because there is only

---

[1] By *system*, we mean an assembly of *components* that operates as a single unit. Boundaries between systems might be arbitrary and adjusted as convenient for analysis. Weapon systems, weapons, and assemblies of components within a program could be examples of systems.

Figure 2.1. Overview of the Framework

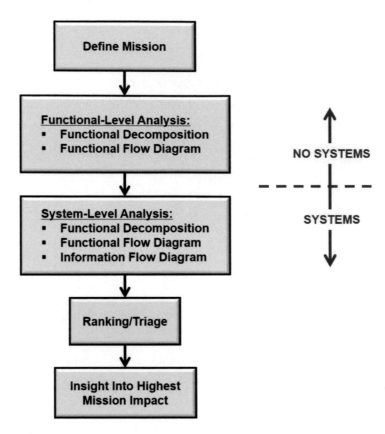

one means to do the data transfer, there is no benefit of distinguishing this mission activity and the systems that perform it. So, some systems will inevitably need to be included at this step, but the emphasis should be on mission activities and not the systems that carry them out.

The next step is to continue the decomposition to include the supporting systems, mapping the functional and information flow among the systems and mission elements. This process follows the same methods used in the mission-level analysis except that it includes systems and extends the flow diagrams to include information flow. Analysis at the system level requires participation of the relevant program offices. Each program office need not examine all systems that support the mission(s) under examination, only those under its purview. What is important to capture are the functional and information flow interfaces to systems outside each program office. In addition, since systems typically support multiple missions and functions, these efforts would reduce the total number of system-level analyses needed to cover all missions. Partitioning workload in this way would economize effort, make the working groups easier to assemble, and cover most systems, since most systems are under program management. Care will be needed to ensure that analysis includes systems not under program management.

A first triage is done based on critical nodes that would interrupt the mission flow. If further pruning of systems is needed, we offer four further criteria for ranking criticality of systems. These are described in detail later in this chapter.

The next stage, after any system-specific pruning, involves a detailed assessment of vulnerabilities and threats on the selected systems of high criticality to the mission. We envision most of the vulnerability work being done within program offices and the test community in coordination with the prime contractor using existing systems engineering artifacts. The list of critical systems would then be used to form production requests for threat assessments from the intelligence community. Other intelligence and counterintelligence collection could also be used to assess the threat to these critical systems and mission elements. The program manager and authorizing official would use these insights to guide which risks might be accepted and which explored for mitigations. Those mitigations might be found in a number of areas: by changing the functional flow logic at the mission level, by changing the concepts of operations at a more detailed level, or by making a change to a system or system interfaces.

Before describing this process in detail, we pause to reflect on how to define the scope of the missions to analyze in the first step.

Defining a Mission

A mission-level analysis can be rendered easier or harder depending on how much care is exercised in defining the bounds of the mission. We set a goal at the outset of developing a framework that can be executed on some reasonable total of Air Force missions at risk of cyber attack. Defining missions too broadly or too narrowly causes distinct problems for the analysis.

Defining Missions Too Broadly

If the missions are defined too broadly, the main concern is that working groups for the analysis will require an enormous range and depth of expertise, and that this range and depth will prove to be impractical to assemble. Any gaps in the expertise of the working group will create gaps in the final analysis. Experience from the CRRA indicates that the team is also likely to become overwhelmed with the scale of the problem and, if it completes the analysis, it is unlikely to achieve the necessary level of detail for sufficient insight.

Ironically, very broadly defined missions can also lead to redundant analysis efforts across missions. Looking at a deliberately overly broad mission definition illustrates how these problems emerge. Consider offensive counter air in its broadest scope. We might rightly consider that this mission includes the various platforms that directly provide the offensive counter air and the weapons they carry for this mission. We might also include the battle management and command and control needed. Our package will generally need aerial refueling support, intelligence support, logistics support, weather prediction, and so on. Without care, we might very well include a large fraction of the capabilities of the entire U.S. Air Force in this one mission and could very well include capabilities of other services as well.

While the integration of the forces, which this view provides, would be insightful if done well, it is not practical to assemble all the requisite expertise in one working group. With this broad mission definition, we have set for ourselves too large a task in one slice. And it somewhat loses the emphasis of offensive counter air in a very large constellation of other activities, most of which exist to support a wide range of other missions, not just this one. If we include a detailed breakdown of battle management for offensive counter air, it does not make much sense to do it again for defensive counter air, and for global precision strike, and so on. We have not only made the job too complicated for one working group but have duplicated work that others will need to perform as well.

Here is another way a mission can be defined too broadly. Suppose that we define the mission as suppression of enemy air defenses and are truly agnostic of how this might be accomplished. We then find ourselves analyzing every conceivable manner of doing that mission. There are many ways that mission might be done: suppression by air-launched missiles, by ground forces, by ground-based missiles, by electronic warfare, by cyber attack, and so on. During an analysis of alternatives, taking such a broad view is a useful approach because the goal is to find the right materiel (or non-materiel) solution to the problem. But our goal is to examine the mission impact of cyber attacks against systems that support these missions. We are eventually going to trace these missions to systems and need to narrow the problem enough to bound the assessment to a coherent set of U.S. Air Force systems. It seems sensible to limit this kind of mission to something like "suppression of enemy air defenses by kinetic means from the air." That leaves the need for comparable missions of "suppression of enemy air defenses by nonkinetic means from the air" and "suppression of enemy air defenses by nonkinetic means through cyberspace." These could all be combined if an effective working group can be assembled that covers the full range of needed expertise.

Defining Missions Too Narrowly

On the other hand, if the missions are defined too narrowly, a very large number of missions will need to be analyzed for them to encompass all relevant U.S. Air Force missions. We just discussed how suppression of enemy air defenses could itself become three mission areas and not include any of the major supporting activities (intelligence, logistics, command and control, etc.). A large number of missions increases the likelihood of both missing key activities in gaps among the missions and duplicating effort in overlap among missions.

To illustrate, consider constructing narrowly defined missions that perform aerial refueling. Suppose, for the simplicity of tasking to a work team, that an aerial refueling mission is defined as one aerial refueling aircraft supplying fuel to one bomber in the course of a nuclear mission. To encompass all aerial refueling, we would also need separate missions: one platform supporting a refueling orbit in a tactical situation and one for ferrying aircraft. And we would need a separate mission for aerial refueling support to special operations. If we decided to keep these simple and restrict the receiving aircraft to be from the U.S. Air Force, we would further need to define separate missions for supporting Navy and North Atlantic Treaty Organization allies.

All of these do, of course, have important differences that might be of interest in certain contexts. A division like the one above would require up to five working groups, each of which would be analyzing missions that have quite a lot of their activities in common. By defining these missions so specifically—in this case, *one* platform providing fuel to *one* receiving platform—there would also be a risk of missing some critical activity, namely, the coordination that happens when refueling platforms work as a team. Aerial refueling, then, should be viewed as airborne off-loading fuel at a given rate in a theater. This view involves multiple tankers and receivers.

Recommendations for Defining Missions

One essential criterion for defining missions is that they *include no systems*. We exclude from consideration that a mission be defined in terms such as to "provide aerial refueling using a KC-46A." This criterion is imperative for the methodology described in this report. Missions should express the activities that the Air Force needs to perform. They are the goals of *what* the Air Force must do; they are not *how* the Air Force will do them. Missions and all the components that comprise them are expressed as *verbs* (activities), not *nouns* (systems). The role of specific systems enters the analysis at a later stage.

As mentioned above, there will be inevitable areas in which systems are somewhat implied. The very definition of providing aerial refueling implies the use of an aircraft of some kind. And a mission described as "suppression of enemy air defenses by kinetic means from the air" implies the use of an aerial platform of some kind. What is important in these examples is to not specify that the aerial platform is specifically the KC-135, the KC-10, the KC-46A, or the MC-130 and leave that specification to a later stage by a different working group. The platform is an option, and the choice of option provides some robustness to the mission.

That an aircraft of some kind is needed for aerial refueling leads to the necessary observation that a key activity will be to generate sorties of aircraft. Without specifying which type of aircraft, a range of combat support functions are implied—fueling, maintenance, supply chain support, medical support, security forces, civil engineering, and so on. These activities can, in general, be defined independently of the specific aircraft and need not be analyzed by the working group for each mission. They can be assessed separately as missions in their own right. This breakdown of Air Force activities conforms to the formal DoD definition of the term *mission*.[2]

Defining missions to include supporting activities captures a set of operational activities that should be sufficiently comprehensive across the Air Force and has minimal overlap. This leads us to the second essential criterion for mission definition—namely, that the sum of the missions provides a comprehensive view of activities that the Air Force is responsible for executing and

---

[2] The DoD defines the *mission* as "the task, together with the purpose, that clearly indicates the action to be taken and the reason therefore." Joint Chiefs of Staff, *DOD Dictionary of Military and Associated Terms*, August 2017.

that are of some significant risk of cyber attack. Defining the best set of missions for analysis will probably take some trial and error. We offer a potential list to start in Table 2.1.

Table 2.1. Possible U.S. Air Force Missions for Cyber Mission Thread Analysis

Group	Mission
Combat support	Aircraft maintenance Fuels Munitions Supply
Conventional theater operations	Air drop Close air support Combat search and rescue Defensive counter air Interdiction ISR/BM Medical operations Offensive counter air Special operations Strategic attack Suppression of enemy air defenses Theater command and control Theater aerial refueling Theater airlift
Cyber operations	Defensive Offensive
Nuclear operations	ICBM ITW/AA NC3 Prime nuclear airlift force Strategic nuclear aerial refueling Strategic nuclear bombing Tactical nuclear bombing
Space operations	Communications Environmental monitoring Launch/range Position, navigation, and timing Satellite operations Space control Space situational awareness Theater missile warning/battlespace characterization
Strategic airlift	Aeromedical evacuation POTUS/DV support Strategic airlift Strategic refueling

NOTES: BM = battle management; ICBM = intercontinental ballistic missile; ISR = intelligence, surveillance, and reconnaissance; ITW/AA = integrated tactical warning/attack assessment; NC3 = nuclear command, control, and communications; and POTUS/DV = president of the United States/distinguished visitor.

The idea is to limit the analysis to mission areas of most concern for cyber attack. Therefore, combat operations are stressed over peacetime activities such as training. The list is not meant to be exhaustive of U.S. Air Force missions. It is presented to illustrate a manageable number of missions to analyze for criticality of systems and covers most of what might be a high-impact cyber attack to a mission. The mission analysis can probably be done most efficiently using a simple modularity. Flying and space missions can be viewed at the highest level as having three basic mission elements that all must occur for a complete mission, as shown in Figure 2.2: generate a sortie, fly and execute the mission, and recover.

Figure 2.2. Modularity of Cyber Mission Threads

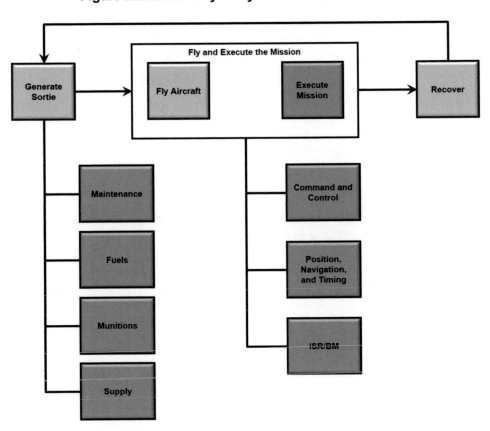

NOTE: BM = battle management and ISR = intelligence, surveillance, and reconnaissance.

The "generate sortie" part of the mission includes tasking the unit, planning and rehearsing the mission, performing maintenance and fuels support, and building and loading any munitions—all supported by a supply system.[3] Many of these activities will be common across flying missions. The four we highlight are maintenance, fuels, munitions, and supply. These could be done once for a flying mission (or perhaps once for a combat aircraft and once for mobility aircraft), archived, and revised as needed for each mission thread. This kind of modularization is one way to economize on effort. Some activities for aircraft preparation will be specific to a given mission and will need to be done for each mission type. For example, only aircraft carrying munitions will need to have the munition building and uploading activities, and a limited number

---

[3] Combat support includes other activities, such as civil engineering and security forces, which also need some form of analysis but most likely need a tailored methodology.

of aircraft (e.g., the intelligence, surveillance, and reconnaissance aircraft) will need to have sensors calibrated.

The next large mission element is to fly and execute the mission. These are the bulk of the flying missions listed in Table 2.1. Two major activities comprise this portion of the mission. The first is the flying of the aircraft, including maintaining situational awareness and avoiding or defeating enemy air defense systems. The second is executing the specific mission, which includes the mission-specific activities associated with the selected mission. The basic flying activities critical to the success of any mission of an aircraft need only be done once, archived, and reviewed for each mission thread. Many of the supporting activities will be part of command and control; position, navigation, and timing; and intelligence, surveillance, and reconnaissance/ battle management. The supporting activities will be used by a mission intermittently or continuously throughout the mission. (Note that we have not listed communications as a separate mission, except for space. That is because most communications systems will arise as systems during the analysis.) These supporting missions could again be done once as a module, archived, and reviewed for each mission thread analysis.

The final and third major mission element is recovery. Because of the inherent circular nature of the mission, aside from the landing of the aircraft, many of the recovery activities could be included in the sortie generation part of the mission for simplicity.

Modularization of this type would then relegate many of the supporting mission elements to a single execution and archive them so that they need not be done repeatedly. The activities of flying an aircraft, too, could be done once and archived. One example of the economies thus realized can be seen with the F-16. The F-16 can support the missions of defensive counter air, offensive counter air, suppression of enemy air defenses, interdiction, close air support, and strategic strike. The sortie generation and flying part of the mission can be done just once and used in common in all these mission threads. Most of the activities for a mission thread would then be concentrated in the mission-specific part of the thread.

One final comment on the missions. There are limits to what an analysis of this kind can reveal about comparisons *among* missions. We discourage any attempts at extending the analysis to a higher level than these missions—for example, by defining a single Air Force mission with the missions in Table 2.1 as mission elements. We see no promise in attempting to analyze, by some metrics, whether, for example, close air support or tactical nuclear bombing is more critical to the Air Force. These comparisons are strategic questions that must be resolved by the judgment and priorities of leaders at the Air Force or national level.

Triage

Mission Threads

Our proposed triage framework begins with standard systems engineering methods to decompose the mission into lower-level, system-agnostic mission elements (functions, activities, etc.) before proceeding to systems. By considering how to disrupt the mission flow, the most critical systems to the mission are identified.[4]

Functional Decomposition

Every mission is composed of various activities, which form the basic elements of our mission impact analysis methodology. To determine the mission elements that comprise a given mission, it must be *decomposed into a functional flow block diagram*, a well-known technique in systems engineering.[5] As suggested by its name, this process of decomposition is functionally, not system or solution, oriented. For example, activities should be phrased as verbs, not nouns; "determine if the runway is clear for landing" is preferable to "runway obstacle detection system." Activities can potentially be implemented in many ways while systems instantiate a particular solution and thus limit the analysis.

The functional decomposition starts with the mission definition and is then repeated at lower and lower levels, ending at the point where it is no longer possible to remain system agnostic. At the highest level of decomposition, the activities could be "mission-essential activities, functions, or tasks," which can sometimes be derived from doctrine, concepts of operations, concepts of employment, or other authoritative sources. At lower levels, these are the subactivities, tasks, functions, and other mission elements that comprise each higher-level activity. The terminology is not as important as their criteria; what is critical is that they are system agnostic and *mutually exclusive and comprehensive* of the mission in question.

To obtain such a set of activities, it is possible to first list all possible activities that go into the mission and then refine the list by combining "similar" or related activities into higher-level activities. The hierarchical decomposition from higher-level to lower-level activities allows for traceability back up to the highest-level activities, which can be useful in informing mission impact.[6]

Two lessons were learned from the experience of mapping a few initial mission threads. The first lesson is that some systems will inevitably need to be included in this part of the analysis, as some mission elements and the systems that instantiate those missions are practically inseparable. The second lesson is that only mission-essential activities should be listed here, not every

---

[4] We want to avoid unnecessary confusion over terminology. Terms such as *mission, function,* and *capability* are often used in a restricted sense in the DoD. There is no need for such distinctions in this context and they force specificity that can confuse the overall concept. We distinguish between activities that need to be done (missions, functions, capabilities, tasks) and the systems instantiated to carry out those activities. Because the activities will be decomposed in a hierarchy, there is some need to distinguish levels in that hierarchy. For these purposes, we use the term *mission elements* when referring to generic decomposed elements of a mission; specific elements that are directly supported by systems are called *functions*.

[5] Department of Defense Systems Management College, *Systems Engineering Fundamentals*, Fort Belvoir, Va.: Defense Acquisition University Press, 2001, pp. 49–50.

[6] Additional guidance for how to decompose activities can be found in N. Viola, S. Corpino, M. Fioriti, and F. Stesina, "Functional Analysis in Systems Engineering: Methodology and Applications," in Boris Cognan, ed., *Systems Engineering: Practice and Theory*, London: IntechOpen, 2012.

conceivable mission task. Checking the weather before a refueler approaches a receiver and mating the refueler and receiver are both mission activities. However, in many circumstances, even if the first mission element could not be performed, the mission would proceed, especially in a combat situation where risks of not refueling the receiver are dire. But the inability to mate causes mission failure. To the best extent possible, only those mission elements whose loss would cause a mission to fail or abort should be included.

The next step is to extend the mission-level decomposition to the systems that support those missions. Expertise from program offices will be indispensable at this stage. It will be important to ensure that systems not under program management are not overlooked at this stage of analysis. The analysis at the system level is much the same methodologically as the analysis at the mission level. The lowest-level mission elements need to be decomposed into the systems that support them and the functional flow and the information flow among the systems and the mission elements mapped. The functional and informational flow is necessary in order to capture control loops and system interactions.[7] This step can use the same methods for decomposition and flow mapping used at the mission level, so we do not describe them further here. This analysis should be able to build off existing systems engineering artifacts done by the program office, the prime contractor, or some combination. The generation of this analysis might fruitfully be fully integrated into those other efforts.

System decomposition should be continued down to the lowest level that is a discrete design element. There is little additional benefit of decomposing a system into subsystems that cannot be altered without redesigning other subsystems. Hence, some decomposition might go to the level of a line-replaceable unit, and others might stop at the level of avionics, if the avionics are so integrated that one subsystem cannot be altered without redesigning other subsystems.

A lesson that was learned during one pilot mission thread analysis is that it is possible to overlook some key systems during the decomposition if they are not managed as systems. Most of the systems that were overlooked were in some sense infrastructural, such as the 1553 data bus on an aircraft. It is essential that special vigilance be applied to ensuring that all such systems are captured.

It is essential, too, to document the relationships among activities within each hierarchical level using a functional flow diagram. While developed to document the flow among functions within a system, this established technique can be adapted to document the flow of activities within a mission, for each decomposition level. Each connection between activities, documented by an arrow as shown in Figure 2.3 for a single level of decomposition, indicates only "function

---

[7] For a similar approach, see Roberto Filippini and Andrés Silva, "A Modeling Framework for the Resilience Analysis of Networked Systems-of-Systems Based on Functional Dependencies," *Reliability Engineering and System Safety*, Vol. 125, 2014, pp. 82–91.

Figure 2.3. Sample Functional Flow Block Diagram

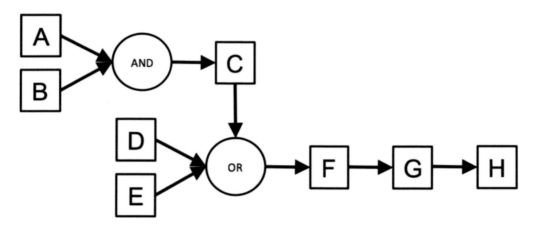

NOTES: This example mission has three paths to completion, with each block designated by a letter representing a mission element: (1) *A&B, C, F, G, H,* or (2) *D, F, G, H,* or (3) *E, F, G, H.* The diagram includes functional sequencing; arrows indicate the order in which activities are completed but not how long they take. Note that this diagram can be extended to include time, with each mission element occupying a representative block of a timeline, but such elaborate diagrams are not needed for our mission impact analysis.

flow and not a lapse in time or intermediate activity."[8] Combinations of activities can also be documented using logical AND and OR symbols, indicating where every path must be followed and where alternative pathways exist, respectively.

Sometimes a mission may be defined as having multiple different goals, such as "locate and characterize surface-to-air missile sites *and* destroy or disable surface-to-air missile sites." In this situation, it is necessary that the functional flow block diagram capture both mission goals, since if either one is not met, the mission fails. The first step in the mission impact criticality analysis is thus to determine whether the goals are dependent on one another. If the goals are independent of each other, and the order in which they are accomplished does not matter, the analysis can proceed with either occurring first (essentially having two functional flow block diagrams, one leading to the next but both necessary for mission completion). If the goals are dependent on each other or if the order does matter, the functional flow block diagram should reflect their dependence and ordering.

Note that functional decomposition is a common feature of many standard systems engineering techniques and standards, including Functional Hazard Analysis.[9] It is also an essential step in more modern approaches to safety engineering based on systems theory, such as the System Theoretic Accident Model and Processes,[10] developed by Nancy Leveson and extended into a

---

[8] Department of Defense Systems Management College, 2001.

[9] SAE International, *Guidelines and Methods for Conducting the Safety Assessment Process on Civil Airborne Systems and Equipment*, Aerospace Recommended Practice ARP4761, December 1996.

[10] Nancy Leveson, "A New Accident Model for Engineering Safer Systems," *Safety Science*, Vol. 42, 2004, pp. 237–270.

formal process called the System Theoretic Process Analysis (STPA).[11] STPA was developed specifically to handle complex systems and emergent failures.[12] Recent work has applied the accident model and analytic approach to security and cybersecurity in what is called STPA for Security (or STPA-Sec).[13] The basic approach of STPA is to (1) identify undesired configurations (accidents and hazardous states of the system); (2) establish the systems engineering foundation for the analysis using control and information flow diagrams; (3) identify potentially unsafe control actions; (4) use the identified unsafe control actions to create safety requirements and constraints; and finally (5) determine how each potentially hazardous control action could occur. STPA and STPA-Sec focus on control and information flow diagrams, which are very similar to (and usually require the definition of) functional flow block diagrams.

First Ranking and Triage

The decomposition of the mission (and high-level systems) and documentation of the functional flows should be done exhaustively for the entire mission. Once done and documented, it is generally useful to understand whether some of the identified mission elements are in some sense more critical than others. The goal is a relative ranking, not a quantifiable rating of criticality against some absolute scale. This information might prove useful for two purposes: to understand the different impacts of system vulnerabilities to various mission elements and to triage which systems to assess in detail if resources for analysis are limited. The remainder of this section on mission-level analysis describes a way to assess mission criticality for these purposes.

Topology and Critical Mission Breaks

Once a mission has been fully functionally decomposed, the functional flow block diagram will include all activities that comprise a given mission, on every possible path from initiation to completion, with minimal systems specified. With a fairly simple transformation this diagram can be used to determine the location of critical mission breaks. For example, consider a very

---

[11] Nancy Leveson, Nicolas Dulac, David Zipkin, Joel Cutcher-Gershenfeld, John Carroll, and Betty Barrett, "Engineering Resilience into Safety-Critical Systems," in Erik Hollnagel, David D. Woods, and Nancy Leveson, eds., *Resilience Engineering: Concepts and Precepts*, Burlington, Vt.: Ashgate, 2006, pp. 95–123; Nancy Leveson, *Engineering a Safer World: Systems Thinking Applied to Safety*, Cambridge, Mass.: MIT Press, 2011; Nancy Leveson, "A Systems Approach to Risk Management Through Leading Safety Indicators," *Reliability Engineering and System Safety*, Vol. 136, 2015, pp. 17–34.

[12] Steven J. Pereira, Grady Lee, and Jeffrey Howard, *A System-Theoretic Hazard Analysis Methodology for a Non-Advocate Safety Assessment of the Ballistic Missile Defense System*, Washington, D.C.: Missile Defense Agency, 2006; M. V. Stringfellow, N. G. Leveson, and B. D. Owens, "Safety-Driven Design for Software Intensive Aerospace and Automotive Systems," *Proceedings of the IEEE*, Vol. 98, No. 4, April 2010, pp. 515–525.

[13] William Young and Nancy G. Leveson, "An Integrated Approach to Safety and Security Based on Systems Theory: Applying a More Powerful New Safety Methodology to Security Risks," *Communications of the ACM*, Vol. 57, No. 2, February 2014, pp. 31–35.

simple, linear mission that consists of four mission elements—*A*, *B*, *C*, and *D*—that must occur in that order from mission initiation to mission completion, as shown in Figure 2.4.

Figure 2.4. Linear Mission Example

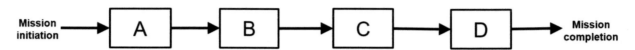

In this linear mission, every mission element from *A* to *D* is equally critical; if any one of them is disrupted, the entire mission will fail. The reason that each mission element is equally critical is that there are no alternative pathways to completion, as there are in Figure 2.5, which shows a slightly more complicated mission.

Figure 2.5. Linear Mission with Alternative Paths Example

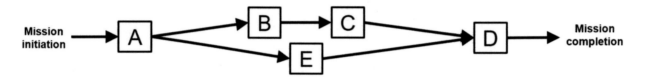

In this example, there are two overlapping sets of activities that could be conducted to complete the mission: (1) *A*, *B*, *C*, and *D* or (2) *A*, *E*, and *D*. It is easy to see that mission elements *A* and *D* are the most critical to the mission because either can cause the mission to fail, while mission elements *B*, *C*, and *E* are less critical (and equal to each other in their criticality), since two paths exist from *A* to *D* and thus from mission initiation to completion.

Note that the key idea in both of these simple examples is to find *single point failures*. The most critical activities to the mission are the ones that, if they alone are disrupted, would cause the mission to fail. Not all missions will have single point failures; some missions are more robust and could require multiple or particular combinations of mission elements to be disrupted in order for the mission to fail. Regardless of whether we are identifying single or multiple point failures, real missions are far more complicated than the simple examples above, and determining critical breaks requires a more sophisticated approach.

Critical Nodes and Paths

A network—or to use the term from mathematics, a *graph*—is composed of vertices (nodes) and edges (links) connecting them. Identifying which vertices and edges are in some sense most critical to the graph is a well-studied problem. There is no universal solution to the problem of vertex criticality, but a number of metrics have been explored that describe various concepts of

19

criticality.[14] Several of these methods can be applied to a functional flow block diagram to identify activities critical to the mission. One method, and the one that we propose for mission criticality, is the *minimum vertex cut set*. This method is particularly amenable to the mission impact analysis as it can be used not only to find the most critical set of activities in a way meaningful to the mission context, but also to create tiers of activities ranked by their criticality to the mission. It is also a relatively simple concept, with established and inexpensive algorithms available for calculating it.[15]

Consider the mission elements as vertices in a graph and the functional flow connections as the directed edges between the vertices. A *vertex cut set* of this mission flow graph is a set of vertices (nodes) that, if removed, will disconnect the vertex that defines the start of the mission from the vertex that defines the end of the mission. In other words, removing the mission elements that form a vertex cut set leaves no path for completion of the mission. The *minimum vertex cut set* is the vertex cut set containing the smallest number of vertices. By a central theorem in graph theory (Menger's Theorem), the number of vertices in the minimum cut set is equal to the number of independent paths between the start and the end of the mission, and hence the number of independent ways the mission can be completed.[16] The size and elements of the vertex cut sets provide measures of the robustness of the mission to functional failures.

Vertices (mission elements) in the mission flow graph that are in the minimum vertex cut sets are most critical to the mission. The vertices (mission elements) in the next largest vertex cut set are the next most critical, and so on. Ranking by vertex cut sets gives a ranking of mission elements by criticality to the functional flow of the mission.

Consider the mission element graph in Figure 2.6 as an illustrative example.

In this notional graph, mission element A is mission initiation and mission element P is mission completion. The mission elements colored red (B, G, H, and I) are all single point failures—they all belong to the minimum vertex cut set of size one. The orange-colored activities (C, D, E, and F) are members of cut sets of size 2 and are the next highest criticality activities. One or more pairs of mission elements must fail to interrupt the mission. There are three such pairs: C and E, D and E, and E and F. These three pairs are the three cuts sets of size 2 and the mission elements they have as members are the second most critical. Similarly, mission elements K, L, M, N, and O shown in yellow are less critical still, being members of cut sets of size 3. Finally, mission element J, in green, is not on a critical path to mission completion and thus is the least critical mission element. Table 2.2 summarizes the cut sets and resulting criticality tiers for this notional example.

---

[14] For a review of a number of criticality measures for vertices and edges in a graph, see M. E. J. Newman, *Networks: An Introduction*, New York: Oxford University Press, 2010.

[15] Newman, 2010; for a review of algorithms, see Frank Kammer and Hanjo Täubig, "Connectivity," in Ulrik Brandes and Thomas Erlebach, eds., *Network Analysis: Methodological Foundations*, Berlin: Springer, 2005, pp. 143–177.

[16] See Béla Bollobás, *Modern Graph Theory*, New York: Springer, 1998.

Figure 2.6. Notional Mission Element Graph

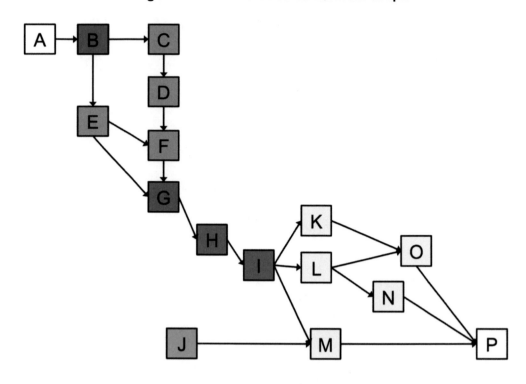

Table 2.2. Cut Sets and Criticality Tiers for a Notional Mission

Cut Set Size	Cut Sets	Criticality Tier	Mission Elements
1	{B}, {G}, {H}, {I}	1	B, G, H, I
2	{C,E}, {D,E}, {F,E}	2	C, D, E, F
3	{K,L,M}, {L,M,O}, {M,N,O}	3	K, L, M, N, O
N/A	N/A	4	J

   If it is necessary to further prioritize within criticality tiers given limited resources, another criterion might be used in addition to the minimum cut set method. The order in which the activities occur in the functional flow diagram gives a slight priority to earlier mission elements because there is potentially more time to recover from failure. Since the functional flow diagram highlights dependencies between activities, those that occur earlier in the diagram could be considered more important from an adversary's point of view. If they attack an early mission element and fail, they might have additional opportunities later. If there are two ways of performing a mission element, or if two systems can perform the same task but one is degraded relative to the other, the vertices can be weighted accordingly in the cut set analysis. Any weighting should reflect the degree of degradation of the performance of a mission element given an alternate system; weighting should not be assigned to estimate mission criticality—that is an outcome of the analysis, not an input.

Functional Flow Block Diagram Transformation

One final matter of detail needs to be explained. In order to apply graph theoretical methods to find critical mission breaks in a functional flow diagram, the diagram must first be transformed into a directed graph (network). A graph does not include Boolean logic to differentiate between alternative paths, but functional flow diagrams do. There is a simple transformation for this purpose. It is assumed that a vertex with multiple incoming edges indicates that each of those edges is an equally valid incoming path—in other words, that they are connected through an OR gate. Thus, the functional flow block diagram needs to be transformed using the merging procedure shown in Figure 2.7.

Figure 2.7. Boolean Logic Transformation from Functional Flow Diagram to Graph

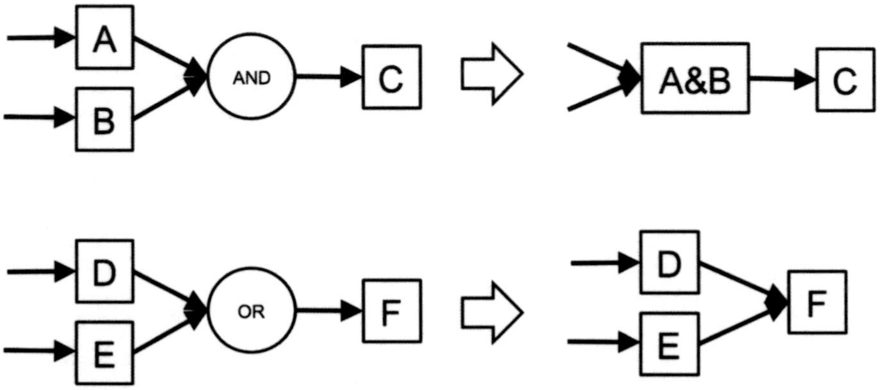

NOTES: The basic transformation from a functional flow block diagram to a graph consists of combining mission elements joined by AND logic and separating activities joined by OR logic. For example, two mission elements, *A* and *B*, that are on paths joined by an AND gate in the diagram, representing two required paths but not a specific order, are replaced by a single mission element block, (*A&B*). Two mission elements, *C* and *D*, that are joined by an OR gate, representing alternative paths, are replaced by two separate paths.

In addition to Boolean logic, functional flow block diagrams can sometimes include conditional logic, indicating where a path is taken if a specific set of criteria is met. While there are a few ways that this type of logic can be incorporated into a graph, the most straightforward approach is to include all conditional paths as alternative paths in the graph. For example, consider the following two paths (where *A* through *F* are mission elements): "Do *A*. If *B*, then do *C* followed by *D*," and "Do *A*. If not *B*, then do *E* followed by *F*," as illustrated in Figure 2.8. The simplest approach is to consider each branch as an equally valid alternative path.

However, this approach only works if the mission elements *C* and *D*, and *E* and *F*, are in fact valid paths to mission completion, regardless of the condition *B* or ~*B* (not *B*). This would be the case if each path is valid but one is preferred for some reason (more efficient, easier to accomplish, etc.). If, on the other hand, elements *C* and *D* can *only* be completed if the condition *B* exists, then they are not on a valid path to mission completion and should not be included in the

Figure 2.8. Conditional Logic Transformation from Functional Flow Diagram to Graph

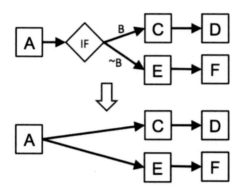

graph.[17] Another possibility is that the condition *B* can be enacted by the operator and thus becomes another mission element within one branch, as illustrated in Figure 2.9. In this case, *B* is under the control of the operator, not an external event such as mission time (day versus night) or weather conditions.

Figure 2.9. Alternative Conditional Logic Transformation from Functional Flow Diagram to Graph

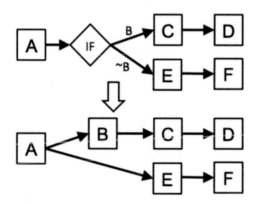

One limitation to this analysis is that using the minimum cut set to differentiate the criticality of various elements does not take into account how often a particular element occurs within a mission. Consider a branch within the functional flow of a mission that is comprised of a single mission element, *A*. This branch is considered equally valid to any parallel branches connected through OR statements; that is, mission element *A* can lead to mission completion just as elements in parallel branches do. However, suppose that mission element *A* is exercised infrequently, say, 0.1 percent of all times that this mission is performed. The minimum cut set approach would likely identify *A* as equally critical to elements in parallel branches, ignoring the difference in frequency. This outcome could potentially misdirect resources to perform a

---

[17] Excluding this path from the graph raises the possibility that the analysis will miss critical elements that only occur under a particular set of conditions. We acknowledge that this is a possible limitation to our approach, but believe that this "scenario dependence" can be captured at the system-specific level, as described in the next section.

system-specific vulnerability and threat analysis of the systems supporting a very low frequency element.

In this situation, there are two approaches to mitigating the limitation. One is to include frequency in the system-specific criticality analysis; this could be done by employing a scenario dependence measure. Another approach is to return to the mission owners, alerting them that there is an alternative path to mission completion that they are exercising infrequently that potentially could be more efficient than the parallel branches. In this case, the analysis would suggest that the element should be exercised more frequently, and that insight should be communicated to the mission owners and operators.

Further Ranking and Triage

Depending on the structure of a mission, cut set analysis might not sufficiently reduce the number of systems for further study. Some additional attributes for triage might be needed. In this section, we nominate four further triage steps.

We make no claim that the criteria described in this section for ranking the criticality of systems to the missions they support are definitive. The four we describe were selected because in our judgment they give meaningful insight into how important a system is to the mission element(s) it supports and can be assessed relatively easily with minimal expertise. They also form a foundation for defining metrics for system robustness to cyber attack that are not especially sensitive to changes in the threat environment. With experience, these might need to be refined and other criteria might be revealed as useful. Some trial and error will be needed as experience is gained.

System Dependencies

This first additional criterion (beyond cut set analysis) is the number of systems and missions that a given system supports.[18] This criterion is a proxy for the centrality of the system in the overall interdependencies of systems and functions that make up missions. A system that supports one critical function and many functions of lesser criticality is generally more critical than a system that supports only one critical function. This criterion is a proxy for that centrality.

Cyber Distance

A second proposal for a system criticality criterion is a proxy for a system's vulnerability to cyber attack.[19] Recall that we defined vulnerability in Chapter 1 as a combination of configurations of a system that can be exploited to harm a mission and access routes to those configurations. Assessing cyber vulnerabilities has been challenging because discovering them is time-intensive

---

[18] This metric is what is called the *out-degree vertex centrality* in graph theory.

[19] Although this introduces vulnerability into the mission impact portion of risk assessment, it provides a useful triaging tool at this stage of the risk assessment.

and requires deep technical knowledge, and the vulnerabilities themselves can change with subtle changes in system configuration. For the purposes of a first-order ranking of vulnerability of systems, we seek a proxy that can be easily assessed with minimal time without deep technical expertise. It is also desirable that this proxy for vulnerability be applicable in the same way across a broad range of technologies so that otherwise dissimilar systems can be ranked using the same criterion.

The provisional proxy that we propose we call the *cyber distance to an unsecure network*. It can be determined in part from the information flow diagram developed in the system decomposition step and is measured by the *minimum number of hops of different protocols between the system of interest and a nonsecure network*. This measure estimates the difficulty an adversary would have in accessing the system arising from the need to reach across dissimilar protocols.

For example, if the system of interest uses operating system *A* and is connected to another system also using operating system *A* and that other system is connected to the Nonsecure Internet Protocol Router Network (NIPRNet), the system has a cyber distance of 0. There are no hops between different protocols before reaching a nonsecure network. If the critical system operates with a protocol different from the one to which it is connected, and which in turn is connected to the internet, it has cyber distance of 1. A nonsecure network such as the internet or NIPRNet has few barriers to the determined and resourced adversary. Each protocol hop that the adversary needs to make to get the system of interest increases the difficulty of access and hence is a proxy for difficulty of cyber access. Systems with cyber distance of 0 are the most critical (vulnerable); system with cyber distance of 1 are the next most critical (vulnerable), and so on.

Even in more comprehensive schemes for assessing vulnerabilities, such as the Common Vulnerability Scoring System (CVSS), the metrics do not explicitly capture hops with changes in protocols.[20] CVSS version 3.0 includes a metric called the *attack vector* metric (part of the exploitability metrics that are in turn part of the base metrics group). The attack vector estimates the distance of the attacker to the vulnerability in terms of whether the attacker needs to be on the same network as the vulnerability, adjacent to the vulnerability, local to the vulnerability, or have physical access to the vulnerability.[21] Because CVSS focuses on more traditional IT systems, it counts hops across network stacks of the same protocols, not hops across different protocols. In a weapon system, the processor controlling a line-replaceable unit might run on a different operating system than test equipment that periodically connects to it. An adversary would need to solve the additional problem of moving malware and executing that malware from one operation system to another, which is possible but would be more work than if the operating systems run on common protocols, and hence is a lesser vulnerability than if these systems ran on the same operating system.

---

[20] FIRST.org, *Common Vulnerability Scoring System Version 3.0: Specification Document*, version 1.7, 2015.

[21] Network, adjacent, local, and physical access are defined more precisely in FIRST.org, 2015.

CVSS is a powerful tool that has become an industry standard for assessing and scoring vulnerabilities. Scoring systems such as CVSS require a lot of detailed information to assess a system and are therefore very useful for the lower-level analysis, after triage, when resources can be directed to specific systems of concern. Yet even at this stage, we caution that there are peculiarities of weapon systems (sometimes called cyber-physical systems in the literature) that differ from networked IT systems, such as protocol hops, that should be part of a vulnerability assessment.[22]

During the pilot cyber mission thread analyses, it was discovered that simple scoring can be useful for cyber distance. The proposed scoring scheme is as follows:[23]

0 = No computers used
1 = Stand-alone machine
2 = Air-gapped machine
3 = Networked, non-internet-protocol
4 = Networked, internet protocol with firewall
5 = Networked, internet protocol without firewall.

This simple 0–5 scoring system is easy to perform and likely good enough for the purpose of triage.

Cyber Separability

The third criterion that we propose for ranking system criticality measures the robustness of system-level support to a mission from a cyber attack. Mission elements (functions) in the lowest level of decomposition identified in the functional-level analysis will be performed by one or more systems. These systems are identified in the system-level analysis decomposition and flow diagrams. One of those systems is the primary system for the function; all other systems are nonprimary to that function and may perform the function in a degraded manner when the primary system fails. For example, a critical function is to stop an aircraft after landing. The primary system for stopping is the brakes on the wheels. Other systems can also stop the aircraft or at least slow it enough. These might include the nonprimary systems of reverse thrust, a parachute, a tailhook, or perhaps a system outside of the aircraft itself, such as a runway arrester. Together, these systems provide a robust capability to stop the aircraft.

Having a suite of systems that can support a critical function results from the logic followed in safety engineering. Safety critical functions are identified and more than one system is designed to support them. Also of importance in designing for safety is that nonprimary systems be highly unlikely to fail at the same time that the primary system fails, which is to say that the failure modes of the systems supporting a safety critical function be uncorrelated.

---

[22] See, for example, Rajeev Alur, *Principles of Cyber-Physical Systems*, Cambridge, Mass.: MIT Press, 2015.

[23] We are grateful to the team at MITRE for devising this scoring scheme.

There are various means for reducing the correlation of failure in these systems. Common means are redundancy and diversity of systems. We need to define these terms precisely. Systems that are identical and that support a common function we call *redundant*. Multiple hydraulic lines and backup hard drives are redundant systems. Systems that are substantially different in design but that can perform overlapping functions we call *diverse*.[24] Brakes, reverse thrust, parachutes, tailhooks, and runway arresters are diverse ways of stopping an aircraft after landing.

For redundant systems, random failures of systems will, in general, be uncorrelated and even if failures are relatively common, reliable systems can be constructed from unreliable parts because the failure probabilities multiply for simultaneous failure. A second strategy that is used is physical separation. If the failure of one system is caused by a physical insult, such as an object striking the system, the more physically separate the systems are, the less likely a common kinetic insult will cause more than one system to fail. Redundant hydraulic lines to an aerodynamic control surface are examples. The redundancy greatly decreases the probability of loss of control for random failures such as a hydraulic leak, and if they are routed separately through the wing, the physical separation reduces the likelihood of failure from a single kinetic insult.

Here is where robustness to cyber attack is different from robustness to random failures and kinetic insults. Redundant systems can provide robustness that contributes to safety from random failures and kinetic insults, but redundant systems do *not* contribute substantially to the robustness against cyber attacks. A cyber attack that works against one system will work as well against another identical system connected to it for redundancy. Furthermore, cyber attacks occur in cyberspace, not physical space, and hence physical separation does not yield robustness to cyber attack. We need a concept for robustness that parallels the concepts used for safety but applies to cyber attacks.

To estimate the robustness of systems to cyber attack, we define two systems as *cyber separable* from one another if no single cyber attack vector could simultaneously degrade the functionality of both. The concept is depicted schematically in Figure 2.10. Cyber separability arises from diversity of systems and a connectivity architecture that sufficiently separates them in cyberspace so as to not be attacked by a single cyber vector. Redundant systems are not cyber separate. Nor are systems that are diverse but depend on a common processor for control or share a common database for operation.

This concept is the one we propose for the criticality criterion. The number of cyber separate systems supporting a function is a measure of the robustness of that function against cyber attacks. Functions supported by only one cyber separable system are more critical than those with greater cyber separability, and that single system supporting it is more critical than systems

---

[24] For a full discussion, see Scott E. Page, *Diversity and Complexity*, Princeton, N.J.: Princeton University Press, 2011.

27

Figure 2.10. Depiction of Cyber Separability of Systems Supporting a Critical Function

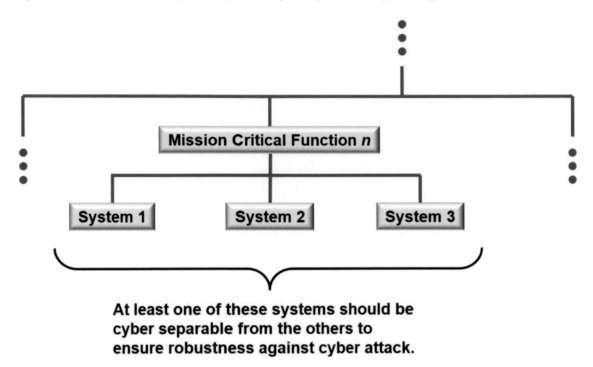

At least one of these systems should be cyber separable from the others to ensure robustness against cyber attack.

that are mutually cyber separable and supporting a critical function. Systems of cyber separability of unity are either the only ones that can perform a function, can perform a function along with other systems but without diversity, or are too directly connected in cyberspace to other diverse systems to be cyber separable. The next most critical functions are those supported by two cyber separable systems, and those systems are second most critical, and so on.

Given that cyber separability has not historically been a design requirement for systems, it might be found that a lot of critical functions are supported by only one cyber separable system. If that is the case, this criticality criterion, although insightful, might not discriminate enough to reduce the analysis space to match analytical resources and other criteria might be needed until design requirements are updated to reflect this concept. An initial scoring might be done in the following fashion:[25]

1 = Fully diverse with two or more cyber separable systems supporting a critical mission element
2 = Diverse but with backup systems being degraded in their ability to support a critical mission element
3 = Redundant but not diverse
4 = No redundancy or diversity.

This scoring was used productively in a pilot mission thread analysis.

---

[25] We are grateful to the team at MITRE for devising this scoring scheme.

Timing

The final criterion that we propose is important for the overall criticality of the failure of a system in supporting a mission, but it is not as easily analyzed as the previous three for all systems and associated missions. However, for some areas, it might be one of the most powerful, as argued for combat support.[26] We list it last for this reason, but list it nonetheless because we feel that in select cases it will provide useful insights not captured in the others. This criterion is one of relative timescales: How quickly (statistically) can the adversary inflict a harmful effect relative to the usefulness of that effect (statistically) for an attack? To introduce this dimension, we turn to a relatively simple example from the past.

During the Second World War, most cryptographic methods used were complicated substitution ciphers. The higher the volume of traffic using such a code, the easier it is for an adversary to infer the encryption key.[27] If all communications—including high-volume tactical battlefield communications—were encrypted at the highest level, the traffic volume would put the encryption at risk for all traffic, including the most critical strategic information. The key insight for mitigating this pitfall was that most information is, to some degree, ephemeral— the value of information to an adversary has a lifetime. For example, a tactical battlefield communication that passes along a command for a unit to move to a new location is of potentially great utility to the adversary before the movement but of little use long after the battle has concluded.

The solution for effectively protecting communications across the spectrum of sensitivity was to use encryption schemes that were deemed good enough that the adversary could not crack the code within the useful lifetime of the encrypted information. That approach reduced the volume of traffic under any one encryption key and allocated the highest-quality encryption to the most dear information with the longest useful life span. The lesson here is that an appropriate measure of risk, in this case how effectively communicated information was protected, was not a function of the difficulty of cracking the code alone. It was a function of the operational importance of the information being encrypted (an effect on an operational mission) and of two dimensions of time: the expected time it would take an adversary to crack the code and the useful life span of that information to the adversary. We offer a criterion for system criticality using a similar frame.

In the case of cybersecurity, the temporal considerations are similar—a time to attack and a time related to the usefulness of that attack. But there is a need to look more deeply into these two time frames to represent cybersecurity risk. Looking first at the time to attack, consider the

---

[26] Don Snyder, George E. Hart, Kristin F. Lynch, and John G. Drew, *Ensuring U.S. Air Force Operations During Cyber Attacks Against Combat Support Systems: Guidance for Where to Focus Mitigation Efforts*, Santa Monica, Calif.: RAND Corporation, RR-620-AF, 2015.

[27] Other factors were important as well, including operational practice and potential compromise of cryptologic materiel. See R. A. Ratcliff, *Delusions of Intelligence: Enigma, Ultra, and the End of Secure Ciphers*, Cambridge: Cambridge University Press, 2006.

sequence of events that constitute an attack. In general, the adversary will need to implant something through cyberspace, which might be done by a variety of means: through a backdoor introduced through the supply chain, through malware installed by spear phishing, or some other vector. The first event, then, is the intelligence preparation and the decision to implant something. The second event is to accomplish the implant. The third is to use or trigger the effect the implant was placed to accomplish. That leads to the fourth event, the effect(s) on the targeted system(s), and the fifth event, the effect(s) on the targeted mission(s).

The key temporal dimensions are the delays between these events, including

- the time for the adversary to implant something
- the time for the adversary to trigger the implant
- the time for the attack to affect a system
- the time for the attack to affect a mission.

Now consider the time frame of the usefulness of an attack. This time frame is a function of how rapidly the victim of a cyber attack can respond and the tempo of the attack relative to the tempo of the targeted mission. The tempo of a targeted mission is an important but subtle point. It is the cyber equivalent of the life span of information in the cryptologic example just discussed. For example, if data are targeted in a data deletion or denial attack and the usefulness or need for that data to support a mission is short, the attack might be ineffectual because its target is ephemeral.

On the response side, there is a parallel set of events with the attack side. There will be a time for detecting and rendering the implant impotent. If this feat can be accomplished, the attack will be foiled. Otherwise, there will be

- the time to diagnose the attack (effect) on the system
- the time to employ a mitigation
- the time for recovery via the mitigation.

The temporal element of the risk, then, is a competition between the attack and response time frames. The overall idea is similar to Boyd's observe, orient, decide, and act (OODA) loop but with a slightly different emphasis. The central idea of the OODA loop is that in a conflict between two adversaries, the side that can observe, orient, decide, and act fastest gains the initiative and keeps the other side confused and ineffectual.[28] Advocates of the OODA loop concept often emphasize the decision times of two adversaries. The concept we posit emphasizes the action times. For a given effect, if the response timeline tightens or the attack timeline expands, the overall effectiveness of the attack diminishes. Likewise, for a given effect, if the

---

[28] The concept of an OODA loop originated with John Boyd. Boyd never formally published his work; for a summary and discussion, see Lawrence Freedman, *Strategy: A History*, New York: Oxford University Press, 2013, pp. 196–201.

response timeline expands or the attack timeline tightens, the overall effectiveness of the attack increases (Figure 2.11).

Figure 2.11. Generalization of the Failure and Recovery of a Capability After an Attack

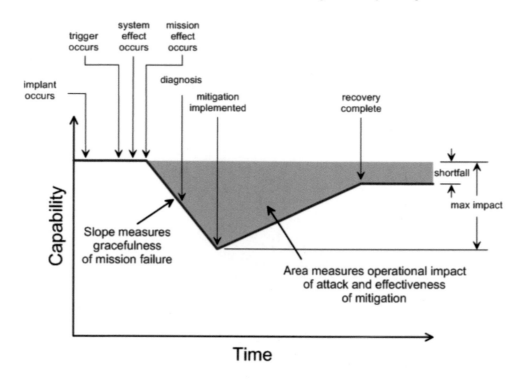

SOURCE: Adapted from Snyder, Hart, et al., 2015, Figure 1.2.

After any triaging, a full analysis of the vulnerabilities and threats to which those vulnerabilities are exposed would be done. This concept can be used to distinguish mission elements of varying importance in the mission flow not captured by the cut set analysis. One way, tried in a pilot mission thread analysis, is described as follows:[29]

1 = Mission impact occurs far in the future after the mission is complete.
2 = Mission impact occurs following the mission, affecting recovery and regeneration of a following mission.
3 = Partial mission impact occurs during the mission.
4 = Immediate impact occurs, causing mission failure or abortion.

Summary

The prototype framework described in this chapter uses a top-down approach of decomposition. A first step decomposes missions into mission elements and systems. The choice of what

---

[29] We are grateful to the team at MITRE for devising this scoring scheme.

constitutes a mission is an important decision in its own right that can render the analysis at the scale of the entire Air Force easier or more difficult. This full mission decomposition and mapping of functional flows is essential in the framework. A method based on a topology of this functional flow can rank the relative criticality of the mission elements and systems. These could be defined to embrace the full spectrum of Air Force missions and, because they would be relatively stable over time, maintained as a library of mission data. The second phase performs four additional triaging steps based on system dependencies, cyber distance, cyber separability, and timing of mission impact.

3. Discussion of the Framework

If the goal is to assess cyber risk to a small set of systems, it might be feasible to study in detail all the vulnerabilities of those systems and to collect and analyze the threats posed both to the systems and the missions they support. Combined with knowledge of how that system supports missions, a fairly complete insight into risks might be known. But when the scale gets much larger than a small set of systems, the number of potential vulnerabilities and threats to examine becomes infeasibly large. A need then exists to identify areas of greatest concern to mission success and, at least initially, limit detailed investigation of vulnerabilities and threats to those most critical areas. Chapter 2 presents a prototype framework for such a triage.

How This Method Addresses the Key Challenges

There is no transcendentally correct method for assessing mission impact or cybersecurity risk. The method we present in this report is one approach developed to satisfy the constraints and desirable characteristics described in Chapter 1. It has been applied in some pilot analyses and will no doubt be further refined. We revisit the desired characteristics to discuss how the framework addresses each one.

The Need to Be Executable at Scale

The principal challenge this framework was conceived to overcome is the ability to be executed at the scale of the U.S. Air Force missions in a reasonable timescale. The framework addresses this challenge in several ways:

- It directs the definitions of missions to be scoped by size and content (system agnostic) so that the missions can be analyzed with a single working group informed by subject-matter experts.
- It economizes on effort across the Air Force enterprise and minimizes redundant efforts across working groups by introducing candidate modular mission elements.
- It uses methods that are familiar to systems engineering and new concepts that do not require complex algorithms that need to be executed by "black box" computer programs created for each mission.
- Most important, it provides a portfolio of criteria for criticality to the mission that can be used for triage and that can be tailored according to available analytical resources.

The Need to Be Updated Throughout the Life Cycle of a System

The prototype framework further facilitates being updated as needed throughout the life cycle of a weapon system in the following ways:

- It separates work that is relatively stable over time from analysis that will need to be updated more frequently.
- It uses proxies for the assessment of mission impact and vulnerabilities that are independent of technological details and hence stable over time as the technologies and threat evolve.

The Complexity of Assessing Criticality in the Face of a Portfolio of Possible Scenarios

A common pitfall of criticality assessments of effects to operational missions is that the criticality is dependent on the scenario in which the operation occurs. Not only are there a wide range of possible scenarios to examine, but they can change over time, further complicating the analysis. The framework presented here extracts criticality from concepts of operations. This focus avoids a direct dependence on scenarios, but still retains a secondary dependence on scenarios to the extent that concepts of operations are scenario dependent.

The Need to Be Transparent and Instill Confidence

The framework uses standard systems engineering methods for mission and system decomposition and the construction of functional and information flow diagrams and does not require complicated, opaque analytical tools.

The Need to Address the Peculiarities of Cyberspace

We identified two peculiarities of cyberspace that deserve special attention in risk assessments: that redundancy does not provide robustness, and that loss of command and control can injure a mission without any system or component failure. The framework embraces these ideas in the following ways:

- A concept of cyber separability is introduced to address the problem of redundancy.
- Functional flow diagrams are incorporated into the mission level and the system level to address the issue of adversary command and control analysis by capturing control loops.

Metrics for Cybersecurity

An additional benefit of this structured way of thinking about mission impact is that it naturally comprises measures for risk. Most of these can be quantified and used as metrics to provide useful insights into risk at both the mission and system levels. All of the criticality criteria serve as potential measures for an existing mission or system that provides insight into how well cybersecurity is being achieved and what impacts might be experienced during attacks against systems or system interfaces.

At the mission level, the vertex cut sets provide metrics on the robustness of the mission architecture and where weaknesses lie. Analyzing these sets should provide insights that might lead to changes in concepts of operations to enhance robustness.

At the system level, four proposed criticality criteria provide insights into how well cybersecurity is currently achieved in a given system:

- Dependencies of systems on one another provide a measure of fragility of the architecture. It is framed in a manner that can be put on contract.
- Cyber distance provides a proxy for how accessible a system is in cyberspace and serves as an approximation of vulnerability. It is framed in a manner that can be put on contract.
- Cyber separability indicates how robust a system is in supporting a critical mission element. It is framed in a manner that can be put on contract.
- The relative (statistical) timing of an attack versus the (statistical) timing of a recovery provides a measure of whether the adversary or the defender has the relative advantage. Although not always easy to assess, when it is possible it provides a powerful insight into the mission impact.

Discussion and Conclusions

The framework presented focuses on cyber attacks, particularly against weapon systems. Whether it is indeed simple enough to be executed across the full Air Force enterprise while also providing enough insight to guide decisions will only be answered by continued testing of the framework on case studies. Most of the information needed to do the analysis already lies somewhere in the Air Force (although some of it might reside with supporting contractors). We have not found existing artifacts from standard processes that are sufficient to supply the necessary information without the use of teams of subject-matter experts.[1] We therefore expect that doing this analysis will require work teams with expertise as broad as the mission is defined and the systems to be assessed.[2]

We see no path to reducing the problem of cybersecurity risk assessment to a turnkey solution executed in a formulaic manner. But this need for assessments by humans has a positive side. We expect that the process of doing the analysis will be as important as the answers it yields. Working groups should gain insights that would otherwise be buried in an intermediate result in a computer simulation program. Furthermore, we expect value in training personnel to think in this kind of structured way.

Parts of this analysis, particularly the mission-level analysis, are of much broader applicability than cybersecurity. Understanding the full mission decompositions and functional flows is essential for a deep comprehension of a mission, for teaching the mission to novices, and for

---

[1] For example, artifacts created by the Joint Capabilities Integration and Development System process, Department of Defense Architecture Framework products, program protection plans, and other artifacts created by program offices.

[2] Mayer et al., 2022, provides guidance for how to use subject-matter experts in the process.

developing future concepts of operations. The decompositions and flow diagrams at the system level can be integrated into existing systems engineering. They should provide crucial insights into risk mitigation and, in addition, into prioritization in budgeting. All of these exercises need to incorporate cyber-specific attributes. Nevertheless, they should be done for the broader understanding of Air Force problems in operations and programming.

There are opportunities to expand and refine this framework. Cybersecurity emerges from a combination of the design of systems and their components (materiel solutions for reducing vulnerabilities) and how those systems are operated (non-materiel solutions). Cybersecurity against offensive attacks is, therefore, not an inherent property of a system, program, or mission, but is a quality of how well the operation of systems in a mission copes with cyber attacks.[3] It is an expression of how well the enterprise responds when under attack. A robust and resilient response to offensive cyber attacks is one for which, even under stress, the enterprise maintains enough resources and command and control over those resources to carry out some minimally acceptable mission accomplishment. To include this full scope, the human interactions with physical systems, both by mistake and as an insider threat, could beneficially be incorporated.[4]

Any approach like this one that attempts to map out how a mission works in order to identify critical nodes and links has inherent limitations. It has been noted that when interconnected infrastructure such as power grids, water supplies, gas pipelines, and transportation networks suffer large-scale damage (e.g., from a hurricane or earthquake), operators sometimes solve problems using novel means.[5] That is, these large systems operate differently under duress than under normal conditions. This problem-solving by operators adds robustness not evident by examining mappings of nominal operating procedures. All of the potential solutions that an operator might use in crisis are impossible to predict in advance. U.S. Air Force missions share this kind of robustness. So the results of this analysis, in this way, reflect a worst case.

We started with two goals in tension with one another—to define a framework for the mission impact portion of cybersecurity risk assessment that is simple enough to execute repeatedly at scale within the Air Force and to yield results that are insightful enough that leaders have confidence to use them in making decisions to accept or to mitigate specific risks. Is it simple enough for enterprise-wide analysis? Is it informative enough to aid in decisions at the authorizing official and system program office levels? To establish that we have adequately met those goals, and to fully verify and validate the framework, the framework will need to be

---

[3] This view of cybersecurity is similar to recent views of system safety. See Erik Hollnagel and David D. Woods, "Epilogue: Resilience Engineering Precepts," in Erik Hollnagel, David D. Woods, and Nancy Leveson, eds., *Resilience Engineering: Concepts and Precepts*, Burlington, Vt.: Ashgate, 2006, pp. 347–358.

[4] James Reason, *Human Error*, New York: Cambridge University Press, 1990.

[5] Emery Roe and Paul R. Schulman, *Reliability and Risk: The Challenge of Managing Interconnected Infrastructures*, Stanford, Calif.: Stanford Business Books, 2016.

applied and tested over a variety of different missions. Potentially beneficial refinements to the framework and methodologies will inevitably be revealed by insights made during that process.

References

Alur, Rajeev, *Principles of Cyber-Physical Systems*, Cambridge, Mass.: MIT Press, 2015.

Aven, Terje, "The Risk Concept—Historical and Recent Development Trends," *Reliability Engineering and System Safety*, Vol. 99, 2012, pp. 33–44.

Bollobás, Béla, *Modern Graph Theory*, New York: Springer, 1998.

Committee on National Security Systems, *Committee on National Security Systems (CNSS) Glossary*, CNSSI No. 4009, April 6, 2015.

Department of Defense, *Cybersecurity*, Instruction 8500.01, March 14, 2014.

Department of Defense Systems Management College, *Systems Engineering Fundamentals*, Fort Belvoir, Va.: Defense Acquisition University Press, 2001, pp. 49–50.

Filippini, Roberto, and Andrés Silva, "A Modeling Framework for the Resilience Analysis of Networked Systems-of-Systems Based on Functional Dependencies," *Reliability Engineering and System Safety*, Vol. 125, 2014, pp. 82–91.

FIRST, *Common Vulnerability Scoring System Version 3.0: Specification Document*, version 1.7, 2015. As of September 14, 2017:
https://www.first.org/cvss/user-guide

Freedman, Lawrence, *Strategy: A History*, New York: Oxford University Press, 2013, pp. 196–201.

Haimes, Yacov Y., "On the Definition of Vulnerabilities in Measuring Risks to Infrastructures," *Risk Analysis*, Vol. 26, No. 2, 2006, pp. 293–296.

———, "On the Complex Definition of Risk: A Systems-Based Approach," *Risk Analysis*, Vol. 29, No. 12, 2009, pp. 1647–1654.

Hollnagel, Erik, and David D. Woods, "Epilogue: Resilience Engineering Precepts," in Erik Hollnagel, David D. Woods, and Nancy Leveson, eds., *Resilience Engineering: Concepts and Precepts*, Burlington, Vt.: Ashgate, 2006, pp. 347–358.

Joint Chiefs of Staff, *DOD Dictionary of Military and Associated Terms*, August 2017.

———, *DOD Dictionary of Military and Associated Terms*, February 2019.

Kammer, Frank, and Hanjo Täubig, "Connectivity," in Ulrik Brandes and Thomas Erlebach, eds., *Network Analysis: Methodological Foundations*, Berlin: Springer, 2005, pp. 143–177.

Leveson, Nancy, "A New Accident Model for Engineering Safer Systems," *Safety Science*, Vol. 42, 2004, pp. 237–270.

———, *Engineering a Safer World: Systems Thinking Applied to Safety*, Cambridge, Mass.: MIT Press, 2011.

———, "A Systems Approach to Risk Management Through Leading Safety Indicators," *Reliability Engineering and System Safety*, Vol. 136, 2015, pp. 17–34.

Leveson, Nancy, Nicolas Dulac, David Zipkin, Joel Cutcher-Gershenfeld, John Carroll, and Betty Barrett, "Engineering Resilience into Safety-Critical Systems," in Erik Hollnagel, David D. Woods, and Nancy Leveson, eds., *Resilience Engineering: Concepts and Precepts*, Burlington, Vt.: Ashgate, 2006, pp. 95–123.

Mayer, Lauren A., Don Snyder, Guy Weichenberg, Danielle C. Tarraf, Jonathan W. Welburn, Suzanne Genc, Myron Hura, and Bernard Fox, *Cyber Mission Thread Analysis: An Implementation Guide for Process Planning and Execution*, Santa Monica, Calif.: RAND Corporation, RR-3188/2-AF, 2022.

Newman, M. E. J., *Networks: An Introduction*, New York: Oxford University Press, 2010.

Page, Scott E., *Diversity and Complexity*, Princeton, N.J.: Princeton University Press, 2011.

Pereira, Steven J., Grady Lee, and Jeffrey Howard, *A System-Theoretic Hazard Analysis Methodology for a Non-Advocate Safety Assessment of the Ballistic Missile Defense System*, Washington, D.C.: Missile Defense Agency, 2006.

Ratcliff, R. A., *Delusions of Intelligence: Enigma, Ultra, and the End of Secure Ciphers*, Cambridge: Cambridge University Press, 2006.

Reason, James, *Human Error*, New York: Cambridge University Press, 1990.

Roe, Emery, and Paul R. Schulman, *Reliability and Risk: The Challenge of Managing Interconnected Infrastructures*, Stanford, Calif.: Stanford Business Books, 2016.

SAE International, *Guidelines and Methods for Conducting the Safety Assessment Process on Civil Airborne Systems and Equipment*, Aerospace Recommended Practice ARP4761, December 1996.

Snyder, Don, George E. Hart, Kristin F. Lynch, and John G. Drew, *Ensuring U.S. Air Force Operations During Cyber Attacks Against Combat Support Systems: Guidance for Where to Focus Mitigation Efforts*, Santa Monica, Calif.: RAND Corporation, RR-620-AF, 2015. As of August 14, 2019:
https://www.rand.org/pubs/research_reports/RR620.html

Snyder, Don, Patrick Mills, Adam C. Resnick, and Brent D. Fulton, *Assessing Capabilities and Risks in Air Force Programming: Framework, Metrics, and Methods*, Santa Monica, Calif.: RAND Corporation, MG-815-AF, 2009. As of August 14, 2019:
https://www.rand.org/pubs/monographs/MG815.html

Snyder, Don, James D. Powers, Elizabeth Bodine-Baron, Bernard Fox, Lauren Kendrick, and Michael H. Powell, *Improving the Cybersecurity of U.S. Air Force Military Systems Throughout Their Life Cycles*, Santa Monica, Calif.: RAND Corporation, RR-1007-AF, 2015. As of August 14, 2019:
https://www.rand.org/pubs/research_reports/RR1007.html

Stephenson, Mark M., *Avionics Cyber Vulnerability Assessment and Mitigation Manual*, Air Force Research Laboratory, March 2014. Not available to the general public.

Stringfellow, M. V., N. G. Leveson, and B. D. Owens, "Safety-Driven Design for Software Intensive Aerospace and Automotive Systems," *Proceedings of the IEEE*, Vol. 98, No. 4, April 2010, pp. 515–525.

Viola, N., S. Corpino, M. Fioriti, and F. Stesina, "Functional Analysis in Systems Engineering: Methodology and Applications," in Boris Cognan, ed., *Systems Engineering: Practice and Theory*, London: IntechOpen, 2012.

Young, William, and Nancy G. Leveson, "An Integrated Approach to Safety and Security Based on Systems Theory: Applying a More Powerful New Safety Methodology to Security Risks," *Communications of the ACM*, Vol. 57, No. 2, February 2014, pp. 31–35.